Boniface Mbunju

Improving Community Water Supply

Boniface Mbunju

Improving Community Water Supply

A case of Tambukareli Sub Community - Geita Tanzania

LAP LAMBERT Academic Publishing

Impressum / Imprint

Bibliografische Information der Deutschen Nationalbibliothek: Die Deutsche Nationalbibliothek verzeichnet diese Publikation in der Deutschen Nationalbibliografie; detaillierte bibliografische Daten sind im Internet über http://dnb.d-nb.de abrufbar.
Alle in diesem Buch genannten Marken und Produktnamen unterliegen warenzeichen-, marken- oder patentrechtlichem Schutz bzw. sind Warenzeichen oder eingetragene Warenzeichen der jeweiligen Inhaber. Die Wiedergabe von Marken, Produktnamen, Gebrauchsnamen, Handelsnamen, Warenbezeichnungen u.s.w. in diesem Werk berechtigt auch ohne besondere Kennzeichnung nicht zu der Annahme, dass solche Namen im Sinne der Warenzeichen- und Markenschutzgesetzgebung als frei zu betrachten wären und daher von jedermann benutzt werden dürften.

Bibliographic information published by the Deutsche Nationalbibliothek: The Deutsche Nationalbibliothek lists this publication in the Deutsche Nationalbibliografie; detailed bibliographic data are available in the Internet at http://dnb.d-nb.de.
Any brand names and product names mentioned in this book are subject to trademark, brand or patent protection and are trademarks or registered trademarks of their respective holders. The use of brand names, product names, common names, trade names, product descriptions etc. even without a particular marking in this works is in no way to be construed to mean that such names may be regarded as unrestricted in respect of trademark and brand protection legislation and could thus be used by anyone.

Coverbild / Cover image: www.ingimage.com

Verlag / Publisher:
LAP LAMBERT Academic Publishing
ist ein Imprint der / is a trademark of
OmniScriptum GmbH & Co. KG
Heinrich-Böcking-Str. 6-8, 66121 Saarbrücken, Deutschland / Germany
Email: info@lap-publishing.com

Herstellung: siehe letzte Seite /
Printed at: see last page
ISBN: 978-3-659-22355-6

Zugl. / Approved by: Manchester, Southern New Hampshire University (2007)

Dedication

This work is dedicated to my wife **Irene Paul Chikira** and our son **Damas Boniface** for their tolerance during my absence throughout the course period.

Acknowledgement

I wish to thank all people who have contributed to the accomplishment of this work. Special thanks should go to Mr. Ligembe N.N who spent most of his time for guidance and supervision of this work. I highly appreciated.

This work might not be completed without kindness and support from the people of Tambukareli sub community. They accepted me to work with them, but the most important here is support they gave me during my studies. Thanks to the organization leaders especially Mr. Hewa who spend most of his time to make sure that the objectives are met.

I wish also to thank my colleagues especially CED class Mwanza center (2007), for their moral and material support during our course period. I would like to express my deep appreciation to Plan International Mwanza and Geita staff for their encouragement and patience throughout the course. Special thanks should go to Plan International Tanzania country management for taking course costs on my behalf as well as giving me time to attend this course.

I would like also to thank Mr. Michel Adjibodou, the director of the Community Economic Development (CED) Programme - Southern New Hampshire University (SNHU) and Mr. Felician Mutasa, Academic Coordinator (Open University of Tanzania), for their contributions in my new skills gained as a result of this work, very much appreciated.

.

List of Acronyms

CED	Community Economic Development
CDP	Community Development Plans
CBO	Community Based Organization
DPs	Distribution Points
DFID	Development Fund for International Development
G G M	Geita Gold Mine
HESAWA	Health Sanitation and Water
HIV/AIDS	Humane Immune Virus/ Acquired Immune Deficiency syndrome
K A S	Knowledge Skills and Abilities
NGO	Non-Governmental Organization
NSGRP	National Strategy for Growth and Reduction of Poverty
PRSP	Poverty Reduction Strategy
MDGs	Millennium Development Goals
MoWLD	Ministry of Water and Livestock Development
SPSS	Scientific Package for Social Sciences
SHUWASA	Shinyanga Urban Water Supply Authority
TWUG	Tambukareli Water User Group
U K	United Kingdom
WSSCC	Water Supply and Sanitation Collaborative Council
WAMMA	Water Aid, Maendeleo ya Jamii, Maji and Afya

List of Tables

List of Figures

Table of contents

CHAPTER ONE

NEEDS ASSESMENT

1

CHAPTER FOUR

PROJECT IMPLEMENTATION

CHAPTER FIVE

MONITORING, EVALUATION AND SUSTAINABILITY

CHAPTER SIX

CONCLUSION AND RECOMMENDATIONS

CHAPTER ONE

COMMUNITY NEEDS ASSESMENT

The purpose of this chapter is to provide evidence on how this project responded to a community's real needs. It is intended to answer the question; what was the real problem and how was it addressed in the community. It also elaborates how communities and other stakeholders in the project area recognized it as a true need and accepted it as their own.

1.1   Community Profile
1.1.1 Location:
The project area is located in sub urban area within the Tambukareli sub community. Tambukareli sub community is among seven (7) sub communities that constitute Ihayabuyaga Village located in Kalangalala Ward in Geita District. The community (Tambukareli) borders with Mkoani sub Community to the North, 14 Kambarage sub community in the south, Moringe sub community in the East & south, and Nyanza sub community in the west.

1.1.2 Population Characteristics.
According to 2002 population and housing census, the community had a total 725 households with an average household size of 4.9. The census report also showed that, Tambukareli community had a population of 1,783 people out of which 926 are females and 856 are males. The district average annual population growth rate including Tambukareli is 3.4%. Children under 18 years as per 2002 population and Housing Census results were 392,675, which is equivalent to 55.4%.

Table 1. Population & Households distribution

Category	Male	Female	Total	Household
Geita District	355,823	356,372	712,192	115,640
Kalangalala Ward	26,853	25,634	52,487	10,652
Tambukareli Sub Community	926	856	1,782	725

Source: 2002 Tanzania Census Report

1.1.3 Social economic and administration of the area

Administratively, the community is under leadership of the chairperson who is selected by the members of the community. Sub community leadership *(Mwenyekiti wa Kitongoji)* has other ten members who form community leadership team. Main occupations of the people living in the area are; petty businesses, agriculture, employment in various sectors of Government offices; self-employment and most are employees of the nearby Geita Gold Mines Company (G.G.M). The Geita District per capita income including the Tambukareli as calculated in 2004 amounts to 152,000/=. Per annum[1]

There is mixture of people from all over the country because the sub community is located in sub urban settlements. Many mining workers are living in this area.

1.1.4 Housing status

The area is predominated by medium cost housing on high-density plots, which have ultimate capacity of 133 people per hectare. Most of the houses have been constructed with burnt bricks and roofed with iron sheets.

6

---

[1] Geita District Planning office

1.1.5 Health facilities and Education:

There is one private owned primary school. There is no health institution within the community. However, those services are obtained in the neighbor Schools while the district hospital is located few kilometers from the community. The majority of the population in Tambukareli sub community depends on local shallow wells for their water needs. Walking for fetching water becomes longer. At this time, pollution of water sources increases dramatically, that leads communities to great risk of water borne diseases.

1.1.6 Water Supply Services:

Most of the residents in the proposed project area depend on local shallow wells of which sometimes get dry during the dry season. However, there is one deep borehole installed with a hand pump. The borehole has a capacity of 9,000 liters per hour. The communities drilled this borehole in 2002, with support from Plan International. The borehole has been a big support to them as it provides water throughout the year. The community had formed a water users group to take care of water supply services.

1.2.0 Needs Assessment

The author visited Tambukareli water users group in October 2005. In his visit, he managed to have focus group discussion with group leaders. The main agenda was to identify major challenges or problems facing them that they thought needed to be solved. The main issue, which was mentioned by the group leaders, was inadequate supply of water services in their community. In order to get the insight of the identified issue, the process went through the structured discussions with community members and other community leaders as well as district officials from related departments e.g. water, health and community development.

7

Some literature from the internet, books and reports were also reviewed as regard to water supply with its effects to the people's health and economic well-being. The process started with random individual and different group interview to get the general feeling of the community on the problem. Various groups like women, men, youths, and children and disabled were involved. Thereafter, a structured discussion was done with leaders of water user group leaders in order to analyze the situation. The group went through in defining the problem and what they thought were the prevailing problems that community would like to solve.

A public meeting was organized and attended by most of the community members. During the general community meeting, group members and community in general listed, prioritized, suggested and agreed to the solutions of various social economic problems that the community of Tambukareli was facing. Results showed that, the first problem in the community was inadequate access to water supply in their community. They suggested to organize themselves with support from other stakeholders to use the available water source to supply water within the community. The survey done in May 2003 by Plan International, an NGO in Kalangalala Ward including Tambukareli community revealed that, only 11 % of populations are in access to water supply throughout the year. Moreover, in the same year, in preparing community development plans (CDP) which was facilitated by district community development department in collaboration with *Plan International Tanzania,* the need for clean and safe water closer to user home ranked high and the first priority. The district health department reports also showed that Diarrhea diseases continue to cause deaths in all age groups,8.6% of outpatient disease cases are due to diarrhea and causes for 6% of childhood deaths.[2] Therefore, communicable diseases continue to cost children's lives due to the use of contaminated water sources.

[2] Geita District Health Report 2004

According to district water department, Tambukareli community like many other communities in the district is facing acute shortage of water for both domestic and economic activities.

1.3.0 Water Requirements at Tambukareli

1.3.1 Water for Domestic requirements

Water consumption as per water Design manual for the category of public tap is 25 litres per person per day. Therefore, a person needs at least 25 liters of water per day as minimum standard.

1.3.2 Public Institutions requirements:

Public institutions include schools, hospitals, Administration offices, Police, churches, Prisons etc.

In our case we have only one primary school in the project area. The water demand for the school will be adopted as shown in the table number 2 below.

Table 2: Water Demand for School

No	TYPE	UNIT	DESIGN	REMARKS
1	Day school.	L/std/d	10 litres	With pit latrine
		L/std/d	25 litres	With W C
2	Boarding School	L/std/d	70 litres	With W C

Source: Geita District Engineer's office

1.3.3 Growth Rate

A recent water supply study undertaken by the then Ministry of Water and livestock development in Geita town adopted growth rate of 6.4 % up to year 2015 and 5% in the year 2025.

Table 2 below shows Domestic water Demand for the interval of 5 years up to the year 2025.

Table 3: Domestic Water Demand

Year	2002	2005	2010	2015	2020	2025
Growth Rate (%)	6.4	6.4	6.4	6.4	5	5
Population	1,782	2,240	3,223	4,562	4,936	6,426
Consumption Rate per day (Litres)	25	25	25	25	25	25
Water Demand (litres)	56,050	67,500	92,075	125,550	134,900	172,150

Source: Geita District Water Engineer's Office

1.3.4 Institutional Water Demand

The only institution present in the area of Tambukareli is a school known *as Aloysius English Medium Primary School* The data concerning water demand for the school is shown in the Table 4 below;

Table 4: Institutional Water Demand; Aloysius Primary School

Year	2005	*2010*	2015	2020	2025
Day student	500	570	640	640	640
Consumption (litres)	10	10	10	10	10
Total Water Demand (liters)	**5,000**	**5,700**	**6,400**	**6,400**	**6,400**

Source: Geita District Water Engineer's Office

1.3.5 Total Water demand

From the above calculations, the total Institutional and Domestic water demand is tabulated below.

10

Table 5: Total Water Demand

Year	2005	*2010*	2015	2020	2025
Domestic water demand (Its)	67,500	92,075	125,550	134,900	172,150
Institutional water Demand (Its)	5,000	5,700	6,400	6,400	6,400
Total water Demand (Its)	**72,500**	**97,775**	**1331,950**	**141,300**	**178,550**

Source: Geita District Water Engineer's Office

1.3.6 Water demand for planning purposes

The table below shows the water requirement aiming at doing Calculation related to future water requirement for planning purposes.

Table 6: Water Demand for Planning

	Year	2005	*2010*	2015	2020	2025
1	Net water Demand (m3/d)	78.0	109	149	158	195
2	Water loss (m3/d)	11.7	16.35	22.35	23.7	29.25
3	Gross water Demand (m3/d)	125	171	182	224	22.4
4	Maximum day demand (m3/d)	108	150	205	218	269
5	Peak hour demand m3/hr.	29	40	55	58	72
6	Storage capacity required % (4)	31.5	43.75	59.85	63.7	78.4

Source: Geita District Water Engineer's Office

1.4.0 Research Methodology

1.4.1 Research Objective

The objective of the research was to collect data and information, which are essential for determining the extent of insufficient water supply, its effects and how this problem can be solved.

11

The purpose was to collect data and information directly from the people about their feelings, ideas and plans in relation to the expected project of *improving access to community water supply* in Tambukareli sub community. The research also aimed at confirming the needs and priorities that were identified by the communities in Tambukareli and other stakeholders during the needs assessment. The feelings, ideas and facts obtained during the survey was for designing the project itself.

1.4.2 Research Questions

The following questions were used as guideline to the research;

(i)   What are the effects of insufficient water supply in Tambukareli sub community?

(ii)   What measures are to be taken to solve the problem?

(iii) What are the barriers of sustainable community water management, and what are the solutions?

1.4.3.0   Characteristics of the Survey

1.4.3.1   Survey instruments

In order to get reliable and valid data for the proposed survey, two methods of data collection were used:

- **Interview:**

The method was used for collection of data at household level and also was applied to ward and village leaders. This method was helpful as the data was collected on the sport as surveyor(s) managed to meet with the household members physically and get their feelings and opinions about the phenomena freely without any interference. The method was also suitable to use to the households because there was no need for the respondents to write and read anywhere. Thus the households were in a position to understand and interpret those questions accordingly.

o *Interview results*

Based on the survey instrument prepared for the household survey, the following data was obtained directly from the household members interviewed: accessibility of water supply in the community i.e. where do they get water, how long does it take to reach the water source and how long does it take for going, collecting water and back home. Other data include the cost of water per family, availability of water throughout the year and, different use of water at the household level.

The research findings section (1.4.9.0), provides details of interview results

- **Self-Administered Questionnaires**

For the district officials such as health officers, community development officers, planning officers, water technicians and engineers, self-administered questionnaires were used. Self-administered questionnaires were designed to allow the respondents to fill in freely. The questions were closed and open ended, that means in some questions respondents were required to provide short answers of yes or no.

o *Self-administered questionnaires results*

Self-administered questionnaire was given to technical personnel from all administrative levels in the district in order to have a technical and administrative feelings and opinions about the surveyed topic. The following data and information were collected from the respondents: the district plan on water and health sector, the role of the district technical staff in supporting water supply issues, general and technical opinion and recommendation about the water solutions in the area. However, the research findings section (1.4.9.0), provides details of general self-administered questionnaires results.

- **Review of secondary data**

Secondary data related to the intended study was reviewed. Secondary data were mainly focused on the records of efforts and plans for water activities in the project area and its surroundings.

13

The experiences in community water managed project were also reviewed in order to know the weaknesses and success that had been observed and can be applied and improved in future. During the secondary data review, a checklist was used as guideline for collection of required data.

o *Review of secondary data results*

The data and information obtained during the secondary data review include: study area population and health statistics. Statistics on health status and water supply coverage in the study was among the data collected. Experience gained from previous water supply project especially HESAWA. Available water source (Borehole) i.e. its capacity, and possibility of supplying water within the community settlements.

Other data obtained during the review of secondary data include, water requirements for institutions and domestic up to the next fifteen years to come. Summary of data collected during the secondary data review have been presented in research findings section (1.4.9.0).

- **Observation:**

Observation method was used to verify physically the current situation of insufficient water supply in Tambukareli sub community. The surveyor managed to visit various water sources used by the community members and confirm distance, time and even water turbidity.

o *Observation results*

Data obtained during the survey through observation include: distance from household settlements to the water sources. Type of water sources were visited, quality of water obtained in various sources was also seen as well as how water is kept at household level. However, the research findings section (1.4.9.0) provides details of general summary of data obtained through observation.

1.4.3.2 Scales and Contents

A total of 20 questions with two types of responses were asked. First the respondents were asked to give their experiences on the different phenomena. For example, the respondents were requested to explain the type of water source in the community. The lists mentioned were like boreholes, shallow wells, ponds and others. This was in the form of closed ended questions. Secondly, they were requested to give their general opinions, experiences, recommendations or comments on the particular phenomena. These were open-ended questions.

All questions were divided into three main scales of 6 -8 questions each. Scales were based on the research question in section (1.4.2). Contents of the research questions were based on the scales used and research questions. For example: in searching for the effect of insufficient water supply, issues like time, distance and type of sources of water were included and asked. Respondents were also asked to recommend solution to overcome insufficient water supply in the community. The questions related to barriers of sustainable water projects as well as how to overcome them in future were also included in the questionnaires.

1.4.4.0 Psychometrics Characteristics
1.4.4.1 Reliability

Intra observer type of reliability was used in order to measure' reproducibility or stability of the data. Intra observer type of the reliability was used where by stability of the responses were measured over time in the same individual respondents. Respondent's responses were monitored in various ways. For example, two related questions could be asked at either the same time or in another time to the same respondent. This helped to check whether there is stable response from the respondents.

15

1.4.4.2   Validity

In assessing the validity of the survey instrument, content method was used. The content method of validity reviews the instrument and data to check how good an item or series of items appear.

Validity of the data assessment was done through giving the instrument to another person for review. Some of the colleagues especially those who are doing research in related topic were in a position to go through the questionnaire and comments were discussed and included.

1.4.4.3   Administration

The survey was administered by four people led by the author. Other members of the team include one community development officer and the other two are health officers. All survey team members have experience of more than ten years in the field. Survey team members specifically administered individual household interview while the leader of the survey (Author) used to administer questionnaire sent to district officials. Training and quality assurance review of the questionnaires was done together by the team of the survey.

1.4.5.0   Relevant literatures and other surveys on the topic

There are number of surveys done worldwide on improving access to water supply. In Tanzania for example, the government in collaboration with various stakeholders like Water Aid Tanzania has conducted various surveys on water supply in relation to poverty reduction. In 2000, the Ministry of water, Water Aid, National bureau of statistics conducted a survey on water and sanitation in Tanzania. The findings showed that, there is need for people to have sufficient water supply.

This is because; Tanzania Poverty Reduction strategy paper (PRSP) recognizes that poverty alleviation will not be achieved without providing every person with access to safe drinking water. (MoWLD 2002).

Plan International Tanzania conducted baseline survey on water supply needs in Geita district in May 2003 and it revealed that only 11% of the communities in surveyed areas are in access to sufficient water supply throughout the year. Other surveys have been done in other countries like United States of America.

In Florida for example, Florida Council of 100 conducted a survey on Florida's Water Supply Management Structure in September 2003. Some of the findings were; water was considered as a public resource that benefits the entire state and supplied by localities and districts are in charge of water management. Some of the recommendations in the survey were; establishment of water commissions, Establishment of water data Centre and establishment of science Advisory Council.

In Midwest, department of geography at southern Illinois University Carbondale in February 2004 did the research on countrywide projections of community water supply in the Midwest, Midwest comprises of six states of Illinois, Indiana, Michigan, Minnesota, Ohio and Wisconsin. The objective of the study was to collect data on water supply infrastructure capacity and compare the aggregates countrywide capacity to projected value of future water use.

The result of this study was mainly to support state and local water supply planning activities, specifically these projections were intended to support the infrastructure decision making process for those small drinking water systems that lack access to other sources of information on the potential changes in future water demands in their service areas.

1.4.6.0   Survey Methods

1.4.6.1   Design

The survey design was cross sectional, the data was collected at a single point in time within the communities of Tambukareli sub community. The design was chosen because it revealed those issues needed by the communities.

The design was also descriptive because the data and information collected was about the feelings, perception and opinions of the people of Tambukareli sub community.

1.4.6.2   Limit on internal and external validity

First, the method used in selection of the samples gave no room for bias and less error. It was an equal opportunity for any community member to participate in the survey. Therefore a sense of bias had been neglected. Secondly, people recommended in the survey such as communities, village leaders, ward and district officials are the ones who are more responsible in the organization.

Therefore, it was guaranteed that the results from the survey would be applied to the targeted community simply because the data and information were collected from the beneficiaries. The survey arrangement was in place including training of the surveyor's assistants. Pilot test was done to 10 people. At this time, it was revealed that some of the questions in the questionnaire were not answered clearly because they (the question) were not well understood.

It happened also that some of the answers from the respondents were covering a number of questions. Surveyors decided to go back to rephrase the question accordingly. Eventually, it went smoothly.

1.4.7.0   Sample

The statistical method was used to determine the sample size. Simple random and purposive sampling was used. Simple random sampling was used in order to have a good or equal chance for everybody in the community to participate in the survey. This method was selected because the nature of the survey required participation of all stakeholders to express their feeling and opinions freely. A total of 38 households, five (5) district officials and **4** Village leaders were included.

1.4.8.0   Data Analysis

- *Quantitative Analysis*

Statistical Package for Social Sciences (SPSS) software version 12.0 was the main tool used during the data analysis.

The SPSS was used to summarize data collected in different phenomena during the survey for the purpose of interpretation and presentation of the results. During the analysis, various areas of the survey questions were analyzed in order to get information required. SPSS was used to summarize data through frequency tables prepared for various variables. For example, a surveyor wanted to know time spent by household in reaching water source. A category of 15 minutes to 1 hour was used to determine time spent through its frequencies and percentage. Type of water source used was also determined by looking at the frequencies and percentages of respondents on various water sources options outlined like ponds, borehole and private shallow wells. Other categories analyzed include time used by the community to reach the water source and get back home. Respondent comments on water quality were also analyzed. The examples below show how the categories were analyzed.

Example 1

Frequency table: Type of water sources

Category	Frequency	Percentage	Valid	Cumulative Percent
Shallow wells	20	52.6	52.6	52.6
Boreholes	15	39.5	39.5	92.1
Ponds	3	7.9	7.9	100.0
Total	38	100	100	

Source: *SPSS*

19

Example 2

Frequency table: Recommended water solution

Category	Frequency	Percentage	Valid	Cumulative Percent
Boreholes	11	28.9	28.9	28.9
Piped scheme	26	68.4	68.4	97.4
Shallow wells	1	2.6	2.6	100.0
Total	38	100	100	

Source: *SPSS*

Analysis from table 1 used to determine which type of water source is mostly used by the respondents. While example 2, used to detect preferred solution to water problem. Descriptive statistical analysis method was used for data presentation. For example, percentages, pie charts and bar charts.

- *Qualitative Analysis*

Qualitative analysis entailed interpreting data collected during the course of qualitative research. In qualitative analysis, both visual and narrative data was analyzed accordingly.

During the data collection, observation was used as one of the method to verify physically the current situation of the insufficient water supply in the study area. ˙ surveyor took time to compare time spent by women in reaching different waᴛᴇ sources and going back home.

It was also a time for surveyor to compare type of water sources mostly used by the communities in the study area. Other areas observed and analyzed include water quality, domestic and business activities e.g. brick making that need water consumption.

The responses through open ended questions, interviews, reports and meetings minutes were analyzed. This was done through coding the main themes coming out from the respondents. For example, one of the codes used was management issue; in here all responses related to the management of the project were listed. Therefore, after the collection of responses related to management issues and other codes used, conclusion issue is reached and the summary is written. In summary, it was written for example, what the respondents say about the sustainability of the project. Or what the respondents say on possible solution to insufficient water supply. In short, during the qualitative data analysis, it was a process of summarizing what was heard and seen from the respondents and come up with the research findings.

1.4.9.0   Research Findings

1.4.9.1   Respondents characteristics

The major respondents of the survey were household members, community leaders and district officials around the project areas. A total of 48 people were involved in the interview include 5 district officials, 4 community leaders and 38 household members. At least 63% of the respondents were females and 36% were males. Most of the respondents were married (47%), followed by singles (34%), separated and widowed 10% and 7% respectively.

The survey also revealed that, most of the respondents had primary education (68%), and very few who had secondary education (15%). Others had adult literacy skills (10%) and non-formal education (5%). All district officials interviewed had either college or university education.

1.4.9.2   Access to Water Supply

1.4.9.2.1   Water Sources and cost

The survey revealed that, there are three types of water sources available and used by the community members for domestic and other economic activities.

There are shallow wells owned by the individual households and mostly are man-made. There is one hand pump deep borehole, which normally many community members use for drinking water, because they believe that the water from this source is safe compared to other sources. Water ponds available are normally used for other activities like construction, brick making and gardening. Despite three sources, water availability is not sufficient to the majority of the community members although most of the sources can provide water throughout the year. According to survey done, 61% of the respondents said, water supply in the study area is insufficient because of inadequate sources to fulfill the need of the communities.

The study also revealed that only 8% of respondents are getting water free of charge. This means 92% of the community members buy water from different sources, whether public or private and normally they pay to an average of 10 Tanzanian shillings per 20 litres. Therefore, following the different sources of water used by the communities are using (Fig.1), it was revealed that 52.6% of the respondents are getting water from private owned shallow wells and 39.5% from public borehole, while 7% of the respondents are getting water from the ponds.

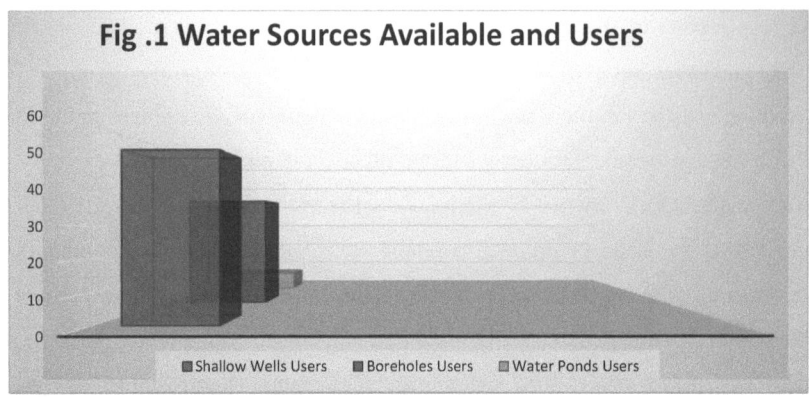

Source: Survey results

1.4.9.2.2   Time and distance in fetching Water

This study recommends total time to fetch water as a more useful indicator than distance to water source. The indicator of total time to fetch water includes going to the water source, waiting, collecting water and returning home. The study revealed that, most of the communities spent an average of 5 hours a day in fetching water. Although the time varies depending on the location of the water source, at least 60% of the respondents said they spend more than 15 minutes to reach the water source. On the other hand they spend more than an hour for going, collecting and returning home. Assume that a mother is collecting water for her family of 6 who need an average of 20 liters a day, if this mother takes a 20 liters bucket and spend an hour to get back home, she needs 6 hours to complete her family water needs. Therefore, distance and time used to reach the water source give a partial indication of the burden of domestic water management felt by women and children in the study area. The time that these women spent in fetching water could be used in other economic or family activities. In addition to that, in most of African countries, children, especially girls, are good supporters of the family in fetching water.

In this case, children will spend most of their time in collecting water instead of concentrating in their studies.

Fig. 2 Total time for collection of Water per Household

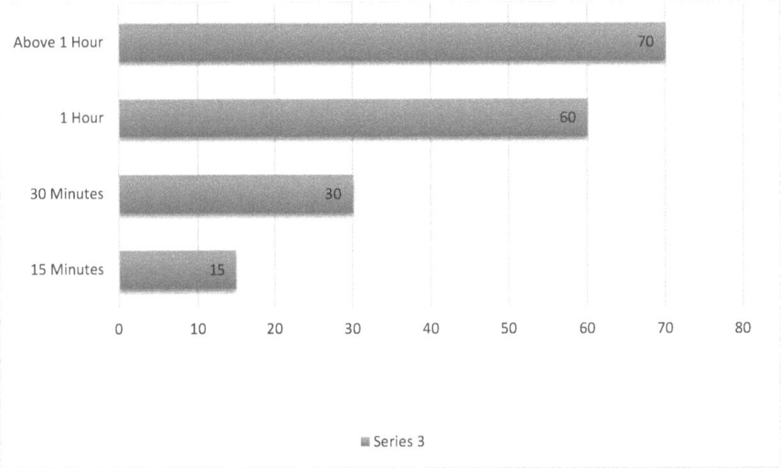

1.4.9.2.3 Water Quality

The study in this area also aimed at getting the feelings of the communities about the quality of water they use. The objective was to know perception of the community members about the quality of water obtained from the three major sources in the community i.e. shallow wells, ponds and boreholes.

About 71.1% of the respondents using water from ponds and shallow wells revealed that the water they use is not clean and safe while 28.9% revealed that, water from the boreholes are safe for human consumption. The analysis showed that, communities of Tambukareli are able to identify at their own capacity the quality of water they use. They know and definitely most of the respondents showed that water from the shallow wells are not safe compared to that from borehole.

1.4.9.2.4   Prevalence of water borne diseases

The study revealed that water borne diseases is at 36%3. According to the district health report, inadequate supply of clean and safe water in the district including Tambukareli area, contributed to the rate of this problem. Therefore, insufficient water supply in the area can cause more health problems if there is no intervention. According to them, the most affected group is children under five.

1.4.9.2.5   Multiple Use of Water

The study also looked into the effect of insufficient water supply in other social economic activities. For example, the study was looking for the main use of water and its current status in the survey area. The survey revealed that the multiple use of water in the area comprises of; domestic i.e. cooking, washing and drinking etc. (76%), gardening and farming (3%), making bricks (13%) and selling (business) (8%).

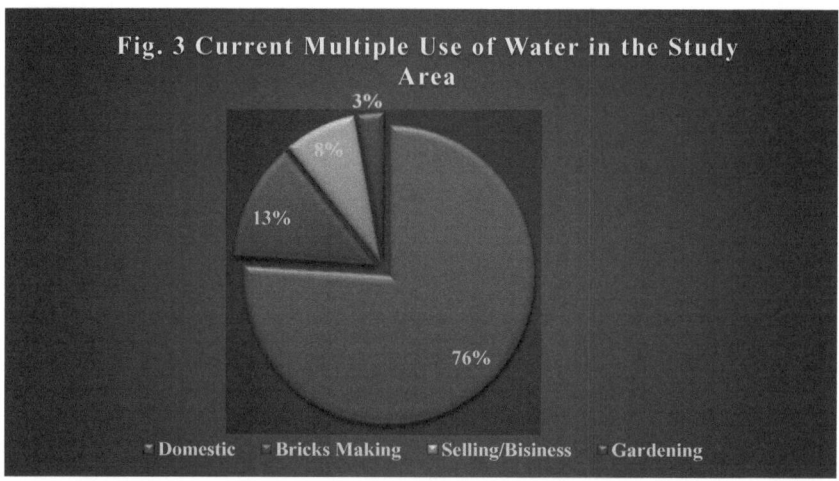

Source: *Survey Results*

The results in current situation of multiple use of water indicate that the available water is mostly used for domestic purposes. They cannot use it for other economic purposes. It is not possible for them to use it for other economic purposes because of its limited accessibility. Some of the respondents interviewed said that, insufficient water supply in the community can make them engage themselves more in economic activities and hence reduce poverty among them.

1.4.10.0   Recommended Options for Sufficient Water Supply

The objective of the study here was to get the opinion and feelings of the communities on what measures should be taken in order to solve the existing water problem. Their feeling was more useful in planning for the suggested option to be taken into consideration.

During the survey, it was learnt that almost 61% of the respondents are aware that there were efforts made by the community leaders and other community members in addressing the water supply problem in their community. However, 40% were not aware. Nevertheless, for the problem solving option, result of the study indicate that 3% of the respondents need more shallow wells to reduce congestion and time for water collection. 29% said their preference was to have more boreholes because they believe that boreholes produce safe and clean water. However, 68% of the respondents said that, the good and preferred option for solution is to have piped scheme because through piped water, they are sure that water will be clean if not safe, but water will be closer to their user home and time spent for fetching water would be reduced.

At least 50% of the people interviewed thought that having their preference option will reduce total time used for fetching water. Additionally, while 37% thought that, having water piped scheme as well as water closer to their homes will reduce distance of water source destination, 13% thought that, this will contribute in reducing prevalence of water borne diseases.

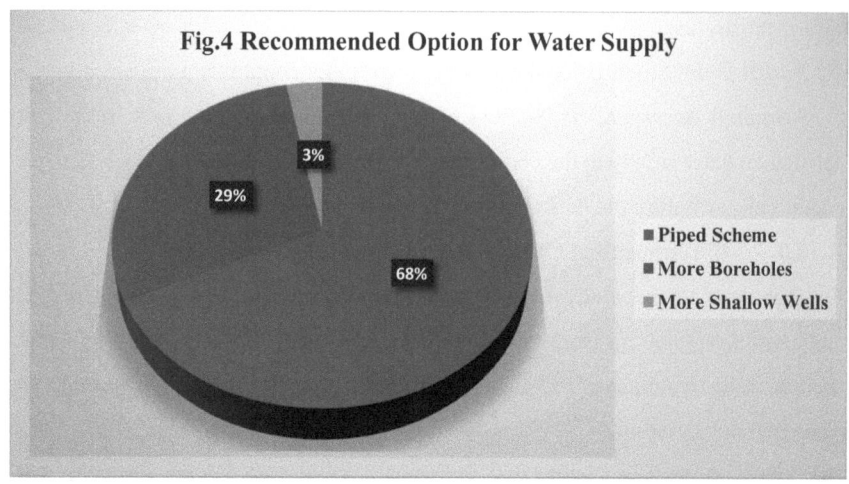

Fig.4 Recommended Option for Water Supply

- Piped Scheme
- More Boreholes
- More Shallow Wells

3%

29%

68%

Source: Survey results

1.4.11.0   Sustainable Water project Management

Sustainability of water supply projects in third world countries including Tanzania has been a permanent agenda. Most of the water supply projects for example in Tanzania, which was constructed after independence in 1970's collapsed due to a number of reasons depending on the area of implementation. Just a few kilometers from the study area, there was a water supply project constructed in early 1970's by the government, but now the project is not working.

The study aimed at knowing why communities thought what could have led to failure of government constructed projects and what were their recommendations for future sustainability of the project. The study revealed that, there are about five (5) major reasons leading to failure of most of the water supply projects.

27

These include; inadequate involvement and participation of project beneficiaries, poor management of funds for the projects, poor contribution in operations and maintenance of the project and poor leadership and transparency especially in water user groups. For sustainability of the intended project, 47% of the respondents believed that good management could contribute better to the sustainability of the project. 21% of the respondent believed that cash contribution would make the project sustainable. While 5% of the respondents said regular maintenance is the best way, 26% of the respondents said all of the above recommendations would lead to the good impact and sustainable water supply project. Therefore, communities at Tambukareli believe that involvement and participation in the project and good management will make their project sustainable.

It was also revealed that for the sustainability of the project, capacity building to the project/organizational leadership should be strengthened.

Communities interviewed however thought that their leaders have no enough capacity to manage the intended project effectively and efficiently. During the survey, 53% of the respondents said that their leaders need more skills and knowledge in order to perform their duties effectively and efficiently.

While 47% said their leaders have acquired capacity to manage their expected project.

Communities and even the leaders have recommended having capacity building sessions to increase their knowledge and skills in leadership and management. The study further revealed that, the intended project cannot be completed unless there is full participation of the communities as well as support from other stakeholders. During the survey, 63% of the respondents said the recommended project need financial support from other stakeholders while 5% of the respondents said management support is highly needed.

However, 32% of the respondents said in order to have a successful project, the intended project need both financial and management support.

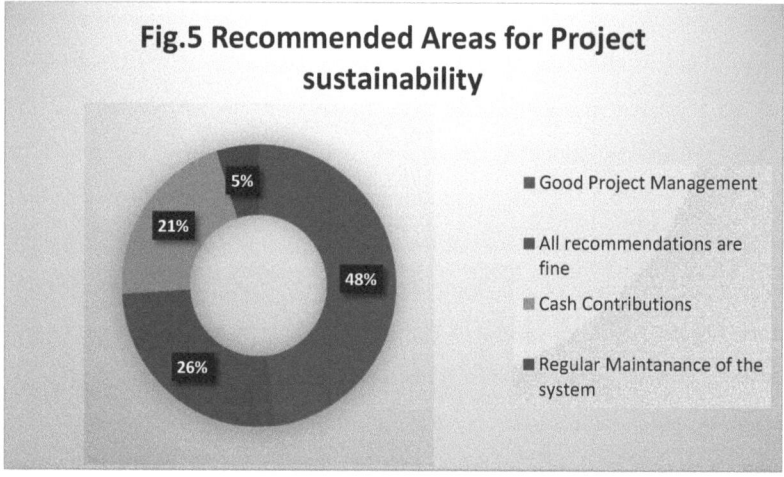

Fig.5 Recommended Areas for Project sustainability

Legend:
- Good Project Management
- All recommendations are fine
- Cash Contributions
- Regular Maintanance of the system

Source: Survey Results

The study also revealed that, communities are ready to participate in the project through various ways. For example, 47% of the respondents said they are ready to contribute in kind, while 13% are ready to contribute in cash. However, 40% of the respondents are ready to participate in the project through contributions in both cash and kind. Generally, 87% of the respondents believed that their colleagues living in Tambukareli sub community would participate to their full capacities to the intended project. Only 13% of the respondents thought that not all community members will participate fully in the intended project. In general, the research result showed that, communities are ready to participate in kind and cash contributions. This is encouraging, and it is highly recommended that project management should make sure that communities are involved in all the processes of decision making in the project.

1.4.12.0   Conclusion Remarks

The survey results tell us that communities at Tambukareli sub community are aware of their problem and they are ready to overcome it. There is need of sufficient water supply is for both social and economic purposes. Communities believe that sufficient water supply in their community would not end for domestic purposes but also it will help them to improve health status as well as increase productivity and hence reduce poverty among them. It is clearly revealed by the respondents that, sustainability of the project will mainly depend on proper management of the project as well as funds that will be generated from the project whether for collection of water tariffs or other contributions (grants from donors and stakeholders). Community involvement and participation could be one among the areas that need more emphasis to create sense of ownership among the members and other community members. Communities believe that good leadership and transparency in the project could contribute to the project sustainability.

CHAPTER TWO
PROBLEM IDENTIFICATION

The objective of this chapter is to define specific area targeted for the change. This includes improving social and economic status of the Tambukareli communities. The current social and economic status was considered low by the community as a result of insufficient water supply in their community. Insufficient water supply was identified as their major problem through needs assessment and research done. During the needs assessment, community's priority was to improve water supply within the community. This chapter highlights the solutions recommended for the change i.e. how it was planned to address the problem. Major stakeholders involved and participated in this project are also identified including their major roles.

2.1 Problem Statement

Only 11% of the populations in Kalangalala Ward including Tambukareli community members are in access to clean and safe water[3]. The burden of traveling long distances in search of water falls on women and children, spending an average of 2 to 5 hours a day in most of the project area[4]. Communities, especially women, spend more time in fetching water instead of carrying out economic activities and thus, increase prevalence of poverty among the families.

The majority of the population of Tambukareli community depends on local shallow wells, for their water needs. Many of these sources are seasonally going dry during the dry season. Towards the end of the dry season water reserves are at their lowest or non-existent.

---

[3] Plan International Tanzania, Baseline survey (May 2003)
[4] Based on community assessment Discussions

Walking for fetching water becomes longer and at this time, pollution of water sources increases dramatically and communities come to great risk of water borne diseases. There is high prevalence of communicable diseases especially in children. Children state they drink dirty water from unimproved wells around the community, which increases the risk of contracting diarrhea diseases.

Diarrhea diseases continue to cause deaths in all age groups. 8.6% of outpatient disease cases are due to diarrhea and cause for 6% of childhood deaths[5]. Therefore, communicable diseases continue to cost children's lives due to the use of contaminated water sources. In general water borne diseases prevalence is at 36%[6]. Communities of Tambukareli would like to solve those problems. Their main focus here first, is to identify the extent these problems affected the well-being of the Tambukareli communities. And second, to find out possible and best ways on how to intervene the stated problem.

2.2   Target Community

The beneficiaries of this project are Tambukareli sub community members and other neighborhood sub communities of Ujamaa and Mkoani. Under the leadership of the water user group, there will be fully community participation in the project implementation. The nature of the project needs some technical know-how during the construction. A contractor will be hired to supervise the construction work and provide technical skills to the communities.

Communities are expected to participate in digging the stretches for pipes fixing, collect available local materials needed such as stones, and sand for construction of delivery points (DPs) within the community settlements.

---

[5] Geita District Health Report 2004
[6] Geita District Health Report 2004

Through the management of water user group, the group members have been trained in the water resources and management of the system.

The aim is to build capacity of the community to manage, run and sustain the system properly. Communities, especially the leaders of the Tambukareli water users group have been trained in operation and maintenance of the system, water tariffs collection skills, simple book keeping and management, environmental protection as well as sanitation and hygiene issues.

2.3 Stakeholders

Different stakeholders involved in this project are as follows:

- Communities are the first stakeholders of this project. Their role is to fully participate through contribution of the resources available such as funds for payment of the local technicians and locally available materials. They should also participate in monitoring, evaluation and management of the project. This will result into smooth implementation of the project, sustainability and improving health and economic well-being of the community.

- Community leaders are very important stakeholders as the ones who link the community and other stakeholders. Their major role is to mobilize and organize the project activities as well as leading the whole process of the project.

- The district Water Engineer provided technical support in water supply related issues to the communities. These include, designing, planning, implementation as well as monitoring and evaluation of the project activities

- Community Development Officers mobilize community, train leaders and organize them. They also participate in planning, monitoring and evaluation of the ongoing activities.

- The health officer provides technical support to the community. This includes training especially in health related issues during the project implementation.
- Plan International and Geita District Council are expected to proceed to provide financial assistance and technical skills during the project implementation. However, Plan International managed to provide funds for the project while Geita District Council provided technical personnel through its water department.
- The role of the author was to support the group in management and administration, design of the project, implementation, monitoring and evaluation of work done.

2.4   Project Goal

The major goal of this project for Tambukareli communities was:

• *To improve health and living standard of the communities.*

The project goal is a dream that communities in Tambukareli are expected to achieve after the end of the project. This is geared to improve social and economic status of the Tambukareli communities. The project is likely to achieve the expected goal because of the current status of implementation.

This will not be a new project to them. Before that, the CBO has been managing the available water source through collection of water tariffs and maintaining all major and minor repair and services. Thus the project fits well with the mission of the organization.

2.5   Project objectives

In order for Tambukareli sub community members to have sustainable water supply closer to their homes as well as improve their living standards, the project need to meet the following major objectives;

• To increase the accessibility of clean and safe water supply closer to the homes of 725 families of Tambukareli sub Community by January 2007
• To build capacity of the community in management of the water supply scheme by January 2007

2.6   Host Organization

The host of this project is Tambukareli water user group. The vision of the organization is:

'Community with improved health and living standard'

The organizational mission is:

'To enable Tambukareli communities to improve their health and economic status through accessible and sustainable clean water closer to their homes by themselves'

The major components that the group facilitated during the project implementation including organization of the project activities, mobilization of the human resources (communities), materials for construction work as well as training of key personnel's who will work in the project.

Other organizations and institutions expected to fully participate in this project include Plan International. This is the NGO which funded this project and expected to fund the project in the near future where necessary. District Council was also expected to fund the project and provide technical assistance through district water engineer's office as well as community development and health departments.

CHAPTER THREE
LITERATURE REVIEW

Various literature on community water supply have been reviewed. The objective was to gather information on similar work done by others, use the information gained from others work and improve the implementation of the intended project. This chapter highlights on theoretical literature which different authors have written on the same activity. Empirical literature reviews how other related projects in various areas in Tanzania were implemented and the lessons learnt from those projects. Finally, the chapter reviews regional and national policies designed to provide framework to this project.

3.1   Theoretical literature

3.1.0   The effect of unsafe water supply

The United Nations states that, 1.1 billion people have no access to water supply (WHO/UNICEF/WSSCC; 2000).

Inadequate access to safe water is at the root of poverty; in whatever way they are defined. They are also both a symptom and cause of poverty. This is due to the following reasons; People's health suffers due to inadequate clean and safe water. In addition, time taken to collect water puts a huge drain on family resources[7]. There is close relationship between poverty and water supply. In other words "water is life". The influence and potential impact of water and sanitation cuts across all the Millennium Development Goals, and making it possibly the most important single area of intervention (Ibid 2005).

The Global Water Supply Assessment 2000 report shows that, the total number of people in Africa with access to water supply has increased considerably over the 1990s.

36

---

[7] Brian Mathew (2005), ensuring sustained beneficial outcomes of water and sanitation

For example, 135 million people in Africa gained access to improved water supply between the years 1990-2000. The majority of these people (87 million) were in urban areas. The report also shows that, the African population is expected to increase by 65% over the next 25 years. This presents a huge challenge to services in the region. To achieve the year 2015-millenium development goal for urban water supply coverage, the percentage of those without access - an additional 210 million people over the next 15 years will have to be provided with service. In rural areas, an estimated additional 194 million people will need to have access to meet the target. Therefore, a total of approximately 400 million additional people will need to be provided with access to improved water supply to meet the 2015 MDG target. Following the findings of the Assessment 2000, this will require a tripling of the rate at which additional people have been gaining access between the years 1990-2000[8]. According to DFID 2004 and WSSCC 2004, the link between water and other Millennium development goals is very clear: The following are some of the millennium development goals:

• *Eradicate extreme poverty and hunger.*

Without water, this goal cannot be achieved to 100 percent. Problems of poverty are inextricably linked with those of water - its availability, its proximity, its quantity and its quality. Improving the access of poor people to water has the potential to make major contribution towards poverty eradication. Provision of more water for agriculture and irrigation will increase food production and help to alleviate the world's hunger.

---

[8] WHO/UNICEF (202) Global Water Supply and Sanitation Assessment 200 report

Improving water infrastructures and services will not only increase water provision, but will also provide jobs to local communities and build capacities. Easy access to clean and safe water will halve the proportion of people who suffer from hunger and whose income is less than $1 a day. [9]

• *Achieve Universal Primary Education;*

Without water supply diarrhea diseases and parasites will reduce attendance and attention of children to schools. Children, especially girls are opt to stay at home to help their parents in fetching water. Easy access to clean and safe water will give girls and boys more time to attend to schools

• *Promote gender equality and empower women:*

Without access to water supply, women spend up to quarter of their time drawing and carrying water often of poor quality. Access to clean and safe water supply will reduce the multiple burdens on women and girls, because they are the primary collectors, providers, users, and managers of water in the household. With their hands free from collecting water, women will have more time to participate in community decision-making and have greater opportunities for livelihood improvement.

• *Reduce Child Mortality:*

Without access to water supply diarrhea disease such as cholera and dysentery will continue killing more than 2 million young children a year[10]. With access to clean and safe water supply, sharp decline in the number of deaths from diarrhea diseases is realized. Better water quality and sanitation services will reduce children's and expectant mothers' susceptibility to diseases, and generally improve health. Better water management will reduce the incidence of waterborne diseases.

---

[9] WSSSCC (2004) source pack on water and sanitation millennium development goals
[10] WSSSCC (2004) source pack on water and sanitation millennium development goals

• *Combat HIV/AIDS, malaria and other diseases:*

Forty million people are now infected with HIV/AIDS, but some countries, such as Brazil, have shown that the tide can be stemmed (WSSCC 2004). People weakened by HIV/AIDS are likely to suffer the most from lack of safe water supply and sanitation, especially since diarrhea and skin diseases are two major common infections

• *Ensure environmental sustainability*

Environmental resources are at ever-greater risk. Some 50% of the entire world's wetlands have been lost since 1900(WSSCC 2004). For example, environmentally sound policies are needed to ensure the sustainability of our ecosystems. One billion people lack access to safe drinking water, 2.4 billion to adequate sanitation. To achieve this target, an additional 1.5 billion people will require access to some form of improved water supply by 2015. That is, an additional 100 million people each year (or 274,000/day) until 2015. (WSSCC 2004).

Therefore, achieving the MDG in drinking water supply coverage will represent a major expenditure in all countries, requiring between US$10 billion and US$30 billion a year on top of the amount already being spent (ibid 2004). Where water problems serve as a constraint to development (e.g. water scarcity, salinity, disasters, etc.), improved water resources management and water supply and sanitation services can facilitate partnerships for global development. The general recommendation of the authors is that, if we are serious in addressing poverty in the world, we need to put more emphasis in the scaling up of water supply to the communities. But the most important thing emphasized is community-based management of the water supply.

Lessons learnt from *Cartel et al (1993)* can be summarized here as follows; 'community managed; through realistic as a medium term measure designed to substitute for the inadequacies of government and private sector suppliers is unproven in long term; Recognition of the need for support mechanism have since been emphasize by *Cartel et all (1999)* and *Schouten and Moriarty (2003)*

There are different international efforts made towards water supply sector. For example; the international drinking water supply and sanitation decade in 1981 of which some lessons learnt were inadequate funding; inappropriate technical solutions and shortcomings of agencies both government and non-governmental.

The Dublin statement on Water and sustainable Development 1992, also addresses: "fresh water is a finite and vulnerable resource, essential to sustain life, development and environment. Women play a central part in the provision, management and safeguarding of water, water has an economic value and should be recognized as an economic good" *(European Commission, 1998).*

Vision 21 was endorsed by the second World Water Forum meeting in the Netherlands at The Hague. The statement was; building on people's energy and creativity at all levels requiring empowerment and building capacity of people; synergy among all partners, encouraging shared commitment among users, politician and professionals within the waiter and sanitation sector, *(WSSCC 2000).*

Despite significant investment in the rural water sector since the early 1970, only 53% of rural population in Tanzania has access to a reliable water supply service and over 30%) of rural water schemes are not functioning properly. Support was provided in a fragmented fashion, and little emphasis was placed on sustainability.

Adequate and sustainable access to safe and clean water plays a critical role in supporting livelihoods and ensuring health in Tanzania.

There are disparities in access to water between urban and rural areas, across regions and even within the districts. The country target by 2010 is to increase access to clean, affordable and safe water from 53% to 65%. (NSGRP 2005).

General conclusion on effect of unsafe and inadequate water supply to social economic development is that, water remains one of the most important resources for human development. This proved the saying that water is life, referred to as; 'there is no life without water'. Various authors mentioned in this section have spelled out the importance of having clean and safe water to the communities. They insist that, the basis of social economic development for any society; need sufficient accessibility of clean and safe water supply.

3.2.0 Empirical Literature

Evidence of the impacts of water supply projects on livelihood, the social cultural life of communities, people's mental and physical well-being, educational opportunities and the like are central to effective poverty reduction strategies both locally and internationally *(Blagborough 2003)*.

One of the hypothesis used by Water Aid stated that; project constructed and managed by the communities have positive impact on the living standards of those communities, particularly in the areas of health, economic status and school attendance. (Blagborough, 2003).

3.2.1 Health through Sanitation and Water (Hesawa)

The HESAWA programme began in 1985 on the basis of Specific Agreement between Tanzania and Sweden in cooperation concerning rural water supply, environmental sanitation and health education. The programme area covers Lake Zone, made up of Kagera, Mara and Mwanza regions, which border Lake Victoria, including Geita district.

The overall aim of this programme was to improve the welfare of the rural population through improved health education, environmental sanitation, drinking water supply, community participation, and capability and capacity building at village and district levels. The pillars or Principles on which HESAWA activities are funded include: Affordability, Sustainability, Replicability, Credibility, and Cost-efficiency.

Since 1985 when the project started, HESAWA produced a marked impact in meeting its objectives, most of which is qualitative, given the nature of the programme. It aimed at creating a qualitative change of attitudes regarding Health through Sanitation and Water (HESAWA) using local resources. To date, HESAWA has undertaken activities in more than 600 villages, spread over more than 180 wards. All 15 districts in the Lake Zone are undertaking HESAWA activities to some degree or other. The programme has expanded significantly during the past year or so. On the basis of village and other studies carried out, the major achievements of the HESAWA Programme can be summarized as follows;

3.2.2 Alleviation of poverty and job Creation

HESAWA provides health education to improve the environment. Improved health indirectly enables people to actively participate in economic activity leading to improved incomes and reduction of poverty. About 1289 new jobs have been created, most private local consultants. With the recent shift in government policy towards higher private sector involvement more jobs are foreseen in the supply of spares and carrying out drilling operations of boreholes.

The government made efforts to promote active participation of the private sector and beneficiaries in service delivery. The aim is to improve efficiency, effectiveness and enhance sustainability of the services.

While water sector is a liberalized area of work for all types of institutions interested, the government through the ministry responsible for water also has the responsibilities and functions of developing, reviewing and further improving water and sanitation policy; facilitating, co-coordinating, monitoring and regulating provision of water and sanitation services to the public with gender perspective; and developing competent sector professionals. The regulatory and institutional framework for water resources management is provided for under the water utilization (Control and Regulation) act. No.42 of 1974, referred to as the principal act as amended in act no.10 of 1981 and written Laws (Miscellaneous) as it is in act. No. 17 of 1989 and general (Regulations) amendment. The act as amended, declares that all water in the country is vested upon the United Republic of Tanzania.

It also sets conditions on the use of water and authorizes the Principal Water Officer with authority, to be responsible for setting policy and allocation of water rights at the national level. For designated water drainage basins with established basin water offices, the responsibilities are under the Basin Water Officer. Currently, a large part of water in Tanzania is used for domestic purposes.

Most of the population (about 80 percent) lives in the rural areas and only the remaining 20 percent lives in urban center[11]. Despite the greater resource potential, many of the sources remain undeveloped. Additionally, a good proportion of the population uses water from undeveloped and crudely developed sources such as lakes, rivers, ponds and shallow and open wells.

By the year 1999 only 45 percent of the rural population and 68 percent of the urban population had access to clean and safe water supply[12].

---

[11] Most C L E A R I N G house, best practices (2002) H E S A W A

[12] Ibid (2002)

While these figures are only national averages, the situation varies a great deal among different geographical locations. With regard to sewerage services, many urban areas continue to be affected by poor sanitary services. Only about 7 percent of the urban dwellers are connected to the existing water piped sewerage system, obtained only in a few towns and where it was constructed a long time ago.

The country's development policy is recommended for putting major emphasis on development of social service sectors, water, included. Not only that, but also the water policy that encourages other stakeholders (other than the government) to engage in water sector development activities is recommended. Today, Tanzania has a long list of institutions both public and private working in the development and delivery of water and sanitation services. There are factors, which denote the existing potentials and opportunities for investment in the water sector for several years to come. This includes greater unexploited water resource potential. There is greater demand for water sector services that are still unmet and still growing.

3.2.3 WAMMA Case Study

During the five years up to March 1996, a collaborative partnership between the Tanzania Government and Water Aid helped a total of 86 communities in Dodoma Region. During the implementation the partnership formed coordination teams known as WAMMA. The acronym WAMMA represented four parties involved in the district team: WA- Water Aid, M-Maendeleo ya Jamii (Community Development Department, M-Maji (Water Department) and A- Afya (Health Department). Of the 6 water projects implemented in 1991-96, 5 were progressively satisfactorily *(Jarman, J and C.Johnson 1997)*. Capacity building was a major important component of WAMMA program. The WAMMA evaluation report showed that the intended outcomes of the project were reached as expected.

For example, at the village level, it is clear that communities have acquired skills such as community organization, running public meetings and bookkeeping. They have also gained experience in project planning and management, resulting in a visible growth in a confidence on their own.

3.2.4 Shinyanga Water Supply project Case study.

In 1998, piped water distribution in Shinyanga was reaching only a third of the town's residents. The rest could walk to buy water from kiosks on the piped network or from one of the town's 16 or so tube wells, or pay even more to get it from vendors selling from handcarts. There were also a handful of community wells and an estimated 1600 private shallow wells in use in the town, mostly unprotected and with water that was usually too salty to drink. These open wells were often close to pit latrines, and were thought to be the prime source of cholera that was killing hundreds of people during drought years. Oxfam responded to an appeal from the regional authorities and worked with local authorities and Non-Governmental Organizations (NGOs) to devise and implement a long-term solution.

Most of the work was done between 2000 and 2003 and the main donor was the UK's Department for International Development. The project has provided clean water and improved sanitation and refuses collection for 35,000 people in seven of Shinyanga town's 13 wards, where life became almost unsustainable during the dry seasons. The town supply has been increased from a maximum of 6000 to 10,000 cubic metres per day. This has reduced the cost of water, and the incidence of cholera and other water-borne infections .A community infrastructure has been built to support delivery and cost recovery for these services. Support staff has been trained in the regional water authority, Shinyanga urban water supply authority (SHUWASA), to carry the work forward into the future. (Oxfam Tanzania).

Lucas Lumali from Ndala talks about the effects insufficient water supply: *"In 1994 the situation was bad. The price of the drought was that water was costing 150-200 Tanzanian Shillings for 20 litres. There was a cholera outbreak in Ndala Ward. Many people died in the outbreak. We dug wells in a valley not far from here, but people scrambled around for water, and the place became very dirty. When the drought returned in 1999-2000, it brought the community together to look for change. It was time something was done to break that cycle of drought and cholera, of having no water. It was Oxfam which has helped us to break that cycle, and give us this water. It has changed the way we live in Ndala at a dry time like this. Before, there was no system supplying water to Ndala."* (Oxfam Tanzania website)

The women can tell you more about water than the men!" said Mwajuma Omary in Ndala. "They do not carry it. They do not either cook or wash up. With water that is salty, one of the problems is that you have to use a lot of soap. Having the water so close to our house makes a big difference to my life. Before, there was no supply in Ndala. We had to buy water. When it was expensive, as it would be at this dry season of the year, we had to go without other things so that we could buy water. We had to use it very carefully, because every cup-full of water cost money. We use 12 buckets of water daily. During that drought season we had only three buckets a day. Now we can wash freely. And I no longer fear the children to become ill. We know Oxfam, because it was Oxfam that brought the water here. Without Oxfam we would have been as we were, without water. (Oxfam Tanzania Website).

3.3.0 Policy Review

Two major policies designed to provide a framework for improving accessibility to Water supply Project reviewed here. However, the Tanzania 2025 Development Vision will be also considered. Some policies related to poverty reduction and water, designed to provide a framework for improving social economic status in sub Saharan Africa and Asia will be also reviewed here.

3.3.1 Tanzania National Water Policy (2002)

The main objective of the revised national water policy is to develop a comprehensive framework of sustainable development and management of the nation's water resources, in which an affective legal and institutional framework for its implementation will be put in place. The policy aims at ensuring that beneficiaries participate fully in planning, construction, operation, maintenance and management of community based domestic water supply schemes. The national water policy of 2002 seeks to address cross- sectorial interests in water, watershed management and intergraded and participatory approaches for water resources planning, development and management.

The structure of the policy contains three major sub sector issues namely; water resources management, rural water supply and urban water supply and sewerage. *(Ministry of Water and Livestock Development- 2002)*. This project is in line with this national Water Policy.

3.3.2 Water and the Tanzania 2025 Development Vision

The Tanzania vision 2025 aims at achieving high quality livelihood for its people to attain good governance through the rule of law and develop a strong and competitive economy. Water is one of the most important agents to enable Tanzania achieve its Development Vision objectives (both social and economic). These objectives are **eradicating poverty, attaining water and food security, Sustaining biodiversity** and **sensitive ecosystems.**

The revised national Water Policy, subsequent reviews and reforms of existing laws, institutional framework structures are aimed at meeting the objectives of this vision *((Ministry of Water and Livestock Development- 2002)*. So, the designed project is geared to attain the Tanzania 2025 development vision.

3.3.3 National Strategy for Growth and Reduction of Poverty (NSGRP)

The National Strategy for Growth and Reduction of poverty, known as MKUKUTA in Swahili *(Mkakati wa Kukuza Uchumi na Kupunguza Umasikini Tanzania),* was approved by the cabinet in February 2005 for implementation within five years. The National Strategy for Growth and Reduction of Poverty established by vision 2025 is committed to the achievement of Millennium Development Goals (MDGs). This strategy has an increased focus on growth and governance. Additionally, it has as well playing roles as an instrument for mobilizing efforts and resources towards its outcomes.

Income poverty status challenges mentioned in the National Strategy for Growth and reduction of Poverty highlights the importance of Water. Improved rural water supply coverage has increased to 53% in 2003. However, 47%) of rural household are using unprotected sources of drinking water[13]. Moreover, a long distance to water sources entails heavy workloads on women and children. One of the National Strategy for Growth and Reduction of Poverty goals and targets is increase access to affordable clean and safe water. Additionally, it has to increase proportion of rural population. With access to clean and safe water from 53% in 2003 to 65% 2009/10 within minutes of time spent on collection of water. Not only that, but also to increase urban population with access to clean and safe water from 73% in 2003 to 90% by *2009/10 (Vice president's office, Poverty Eradication 2005).*

48

---

[13] Tanzania National Strategy for Growth and Reduction of poverty (2005)

Therefore, following the above policies, the project is within the efforts of the Tanzania Government in economic development of the entire communities as well as poverty eradication. Improved access to community water supply is among those efforts.

3.3.4.1.1   Poverty Reduction Strategies Papers and Water in sub Saharan Africa and Asia.

Water supply and sanitation are critical factors in a day to day problems faced by the poor in developing countries. The extent and significance of water related poverty was recognized at the International Freshwater Conference held in Bonn in December 2001, which related the importance of achieving safe, affordable and sustainable water and sanitation access for poor populations, as a central global concern of poverty reduction. Despite the accepted importance of water supply concerns, preliminary analysis of emerging Poverty Reduction Strategy Papers (PRSP) in sub Saharan Africa in 2001 indicated that these concerns have not been adequately reflected. (ODI, 2002).

• **Water, Poverty and Sustainable Livelihoods** The summary of Overseas Development Instituted (ODI) on water policy program in sub Saharan Africa showed that, if water related interventions for poverty reduction are to be meaningful, water objectives in PRSPs need to take account of water resources management as well as water supply and sanitation priorities. Improving access to water supply and sanitation is, of course, not just about taps and latrines: it is about the people and institutions who use and manage them. The water sector has been dominated for many years by perspective emphasizing the health impacts of improved water supply and sanitation.

Sustainable livelihood analysis requires interveners to take a more holistic view of the role of water in support of livelihood activities of the poor. This demands a broader understanding of factors affecting availability, access and use of water as a productive asset and how it is combined with other assets not only to sustain life directly, but also to bring in the income, financial and non-financial, to sustain livelihoods.[14]

• **Towards better integration of water and sanitation in PRSP in sub Saharan Africa - Lessons from Uganda, Malawi and Zambia.** Malawi, Zambia and Uganda share many common features of the status of water supply, health and education delivery as well as some significant differences. In Uganda safe water provision has higher budgetary priority than in Malawi and Zambia, benefiting from a substantial proportion of heavily indebted poor countries (HIPC). In Zambia, civil society succeeded in rising the profile of water and sanitation through the PRSP process, ensuring high visibility in the PRSP document. However, this success was not accompanied by increased budget priority, or better alignment of the sector towards the PRSP, indicating lower political commitment to the PRSP.

Despite Uganda's success in alignment and prioritization of water supply within the PRSP and budget, the water supply reforms in Uganda have not yet to yield substantial improvements in efficiency and effectiveness, although implementation has been scaled up country wide. There is therefore substantial potential for improvement, and sectorial review processes are beginning to grapple with the issue of value for money in the sector.

---

[14]ODI & Water Aid -Water policy Brief 3: July 2002

• **Pakistan new water policy.** For the first time in almost six decades, Pakistan has put together two major policies related with water use and conservation.

The main goal of the water policy is to assure safe drinking water to all at an affordable cost in an equitable, efficient and sustainable manner and to reduce mortality and morbidity caused by water borne diseases.

The policy draft however has yet to satisfy all consumers' rights advocates who say there are many issues such as rehabilitation of dysfunctional schemes inequalities in access, modes of levying of user charges and locations where filtration plants are to be installed. The consumer rights commission of Pakistan (CRCP), an NGO, says every fifth Pakistan child under the age of five suffers from water borne disease and that any new policy should be able to change the situation. CRCP adds that roughly 50 percent of mortality and 20 to 40 percent of hospital admissions are also caused by water borne diseases[15].

The water policy introduces the ides of rising user fees for cost recovery but stops short of privatizing water supply triggering another debate between advocates and supporters of privatization. The centrally formulated water policy makes provincial governments responsible for the service through special agencies that would be created in the cities and districts sub divisions.

---

[15] ODI & Water and Sanitation Program-Water Policy Briefing, No. 5 November 2004

CHAPTER FOUR
PROJECT IMPLEMENTATION

This chapter provides both original plan and the actual implementation of the project. This includes major task and activities undertaken, resources needed and used as well as responsible personnel in each activity planned. It also reports what was accomplished and what was not and the reasons. Generally, this chapter provides a summary of what so far has been done in terms of intended project objectives.

4.1.0 Project Products & Outputs

The expected product and output of this project was a completed project proposal which aimed at sourcing the funds for the purpose of attaining the Communities' objectives of having water supply services closer to their homes. From the initial stage, the project aimed at accomplishing the following activities:

- Preparation and submission of the water supply project proposal to the donors in order to raise funds for the intended project.
- Capacity building; training of water users committee members on water management & water supply scheme management skills
- Construction of the water supply scheme within the community settlements with at least 5 public Distribution Points (DPs)
- Therefore, the major expected output at the end of this project is to have completed and functional water scheme as well as effective and efficient scheme management team.

4.2.0 Project planned Activities.

The Implementation of the project was based on the different planned activities according to activities implementation chart developed.

Activities planned for implementation include; conduct preparation meetings for the project where by different meetings were done including that with water user committees and the community's i.e. public meetings. All meetings were aimed at communicating with the communities and seek their general opinions about the issues affecting them. It was also a process of conducting needs assessment. Completion of the project proposal was also an item that was planned to be completed. During the needs assessment, communities and their leaders showed that they are facing inadequate water supply services in their community and thus they wanted to solve. However, their main problem was how to prepare proposal for fund raising in order to meet their goal. Therefore, a consultant who is the author of this work was asked to support in preparation of the proposals as well as the whole process of securing funds from the donor by that time. *Plan International* Tanzania showed interest to support the idea of the communities in Tambukareli. It was observed that the Tambukareli Water User group was not a registered organization, but was just known by the authorities in the district. They were in the process for registration but they were supposed to have the group constitution. Preparation of the group constitution as well as registration process was among the activities planned for the period of the course and the project.

Sustainability of the intended outputs was also considered, and the capacity of the leaders in operations and management were highly needed. Therefore, training of the water user's group leaders was planned.

The aim was to build capacity of the leaders in water supply management and operations skills. Construction of the scheme after securing funds was regarded as final activity to meet the objectives of the project. It was expected that the project would hire technical expertise for the construction especially in supervision of skilled needed works, while communities would participate in non-skilled works depending on the designed actual construction plan. Generally, the whole process of project implementation was expected to involve different people with different skills. These include; communities, their leaders, community development officers, Health officers, District Water Engineer, the author and other people who were considered to have required skills to achieve the goals and objectives of the project

4.3.0 Actual Project Implementation

The project implementation process started as planned as per implementation schedule prepared.

4.3.1 *Conduct preparations meetings with communities and leaders*

From the initial stage of the project, the author managed to meet with water user group leaders. It was the time where needs assessment was done. The author wrote a letter to the organization leaders requesting to work with the organization and the group leaders accepted the request. The student was later contracted to carry out the tasks of which he had to accomplish during his service period to the organization. During the meeting, various issues were discussed including the successes that the group had; problems and challenges they were facing and the group future development plans. It was during this discussion, that the need for the author to support preparation of the project proposal was developed. Other follow up meetings with the leaders were just monitoring of what had been agreed on and evaluating what was completed as well as thinking about the way forward for improvement. Three public meetings were convened in which there was a very good attendance.

The first meeting was used mainly for introduction to the author as well as to get a rough idea on issues prevailing within the community. The second meeting was for needs assessment, whereby communities made prioritization of various issues they thought were the main development challenges in the community. The last meeting was for finalizing their priorities and selection of other group leaders. New chairperson, secretary and five members were selected to build up a group leadership team.

4.3.2 Project Proposal Writing and Submission

With support from Geita District Water Engineer and other extension officers i.e. community development and health officers, a project proposal was prepared and submitted to the donor, Plan International Tanzania, Geita branch (Appendix 2). After some discussion between the community and other officials, the donor accepted the proposal and later funds equivalent to Tshs. 78,518,300.00 were approved and disbursed for project implementation.

In memorandum of understanding between the community and *Plan International,* it was clearly agreed that, management of funds was to be done by *Plan International* while the community was responsible for materials management and collection of all required materials; especially all locally available materials such as sand, stones, digging the trenches, depending on the design of the project implementation, recommendation from the contactor, and the project supervisors i.e. District Water Engineer and Plan Water Consultant.

4.3.3 Complete Group Constitution and submit to authorities for registration

With the support from community development department, water user's group constitution was prepared and completed. The constitution preparation was done through consultation with other people including other water user groups within and outside Geita district. The registration application attached with the constitution is under process to the registration authorities.

4.3.4 Training of Water Committee on Water Sources and Water Schemes Management

Capacity building through training was done to water users committee on water resources and system management. Before the training, a training needs assessment was done to the organization & leaders. In training needs assessment, various items were looked at. These include organizational analysis, requirements analysis, tasks, skills and knowledge and person analysis.

The needs assessment components consist of the collection of the important tasks, knowledge, skills and abilities (KSA) necessary to perform the job. These items were assessed in order to give inputs in designing the training program. The training was designed to impart the committee members with various skills and knowledge, which will be helpful for the project sustainability. The training contents include; simple book keeping and accounting, water tariffs, project rules, operation and maintenance, system operations, legal and institutional framework. The team of trainers from water, health and community development departments facilitated the training.

4.3.5 Construction of the scheme

Construction of the scheme was expected to start immediately after all required preparation completed. The project construction preparation activities are under Geita District Water Engineer. Moreover, according to memorandum of understanding, this preparation also involved *Plan International* water consultant. So far, various types of pipes and associated materials for the project are already purchased and delivered to the site. It was expected that the construction activity would start in April 2007.

CHAPTER FIVE

MONITORING, EVALUATION AND SUSTAINABILITY

This chapter explains how data and informations were gathered and analyzed in order to anticipate problems, formulates solutions, and evaluates program performance. It elaborates how progresses were measured during the different periods of project implementation. It also provides status of the project changes realized so far and the lessons learnt as well as how the communities and other stakeholders plan for next steps to reach their intended project objectives.

5.1.0 Monitoring

The monitoring of the project was based on the log frame and monitoring and evaluation frame work developed (Table 7) as well as work plan (Appendix 5). Monitoring and evaluation framework contained both qualitative and quantitative verifiable indicators and means of verification. Day to day data collection was done accordingly and the information was analyzed based on the requirement. For example, monitoring tool was developed based on work plan and two main project objectives as follows: Indicators used for the first objective; to *increase the accessibility of clean and safe water supply to 725 families of Tambukareli sub Community by March 2007* were:

- Number of meetings conducted and the participants
- Issues Discussed and agreed upon in the meetings
- Completed proposal and submitted to Stakeholders for funding
- Number of hired contractors and technicians
- Number of schemes including distribution points (DPs) constructed and completed
- Number of family's access to water supply.

Indicators used for the second objective of *building capacity of community in management of the water supply scheme by March 2007* were:

• Number of communities and leaders trained in project management

• The contents/topics of the training covered

• Participant's feeling about the training i.e. understanding, validity etc.

• Completed constitution and approved by authorities

• Registration of the CBO

5.1.1 Research Methodology

The objective of the research during monitoring was to collect data that was used to assess the progress of the project and take appropriate measures if necessary. The following methods of data collection were used for monitoring:

• Interview

The method was used to collect data at individual level especially all leaders and communities who attended the training and public meetings. The method was good as the data collector was able to get feelings and ideas of the respondents directly as there was no need for respondents to write.

The method also was used to training facilitators to get their opinion on how the training went. A monitoring checklist was used during the interview. During the interview respondents were asked to respond to questions prepared specifically in order to get their understanding on the progress of the project. Such as what actually the respondent knows about the project and what he or she knows about the progress of the project so far. It was also a time to gather opinion of the respondents on the progress and ask them to recommend possible ways to improve the ongoing activities for the sake of achieving the intended objectives of the project.

• Review of secondary data

The method was used in order to know what communities have done and recorded. Review of meeting minutes and training reports was done in order to gather data on issues discussed and decisions reached during the community and leaders meetings as well as training done.

The method was necessary not only to know what records were kept by the communities but also it was used to assess how project progress is documented as well as the quality of the records kept. It was a time where weaknesses observed were corrected and appropriate information management system based on the information required was designed.

Data collected were from various sources, this included water user group files, government department files as well as donor files.

Data collected was mainly reports on the progress of the project including challenges and recommendations for improvement. Data was then recorded in the note books for analysis.

• Observation

Observation method was used to check physically the current situation of the project. It was easy for the data collector to visit different proposed areas for the project implementation to check what was going on as planned.

Because monitoring was done in a participatory way, this method was also used as lesson to the communities especially project leaders.

Members in a monitoring team were able to see, discuss and recommend actions to be taken where necessary. It was easy for the project leaders to reach consensus during the discussions, as everybody was aware of what was going on for that particular time.

The monitoring team represented by members of the water user group, officials from health, community development and water departments. Observation involved visiting different sites intended for project implementation. During the visit, notes were carried down to document what had seen in place according to what had been planned and expected at that particular time. After the visit, members took at least one hour to make reflection of what was seen during the visit and come up with a summary. The summary was documented by CED student. Later on monitoring report was prepared after compilation of other data obtained through other means used.

5.1.2 Data analysis and findings

Data was collected based on the monitoring and evaluation framework prepared. Data and information on activities done as compared to what was planned was recorded in the notebooks. However, later, CED student used computer to keep all the data and information gained during the monitoring process.

During the data analysis, CED student mostly used to compare different responses and information gathered in different sources like interview and secondary data. The data was used to check whether the intended activities were going on well as planned or not. It was also used to check any challenges encountered and what action should be taken to overcome those challenges. For example, during the analysis of the data, it was observed that purchase of project materials such as pipes and accessories delayed at least for two months from the expected time of delivery. It was later recommended that, implementation plan should be revised to accommodate the changes expected as a result of that delay. Furthermore, it was also recommended to analyze the consequences that might happen due to the delay of the implementation. The possible consequence predicted was difficulties for implementation if the rainy season would start as usual.

Word processing of the data was applied especially for analysis of qualitative data. A progress report was prepared based on the data gathered and conclusion reached. Monitoring team members then shared it and when approved, it was kept in the appropriate file.

5.2.0 Evaluation

The evaluation of the project was designed objectively to assess the extent to which goal and objectives of the projects have been achieved. Performance indicators outlined in the monitoring and evaluation framework were used as check list during the evaluation of the project. The project was expected to end in January 2007. However, by the end of January, the project was not fully completed as expected. Therefore, this evaluation was meant to assess the extent to which the project objectives were achieved by the end of estimated project period. The evaluation was necessary so as to come out with recommendations and the way forward for the future of the project. In doing so, instead of going deep looking into the relevance, efficiency and effectiveness of the project, a total of three key evaluation questions were used as follows:

• Did we do what we intended to?

• What did we learn about what was implemented and what was not implemented?

• What do we plan to do with evaluation findings for continuous learning?

5.2.1 Research Methodology

The objective of the research during evaluation was to collect data that would be used to assess the extent to which the intended goal and objectives of the project were achieved.

Participatory evaluation was applied, whereby different stakeholders who participated in the project implementation were involved. The following methods of data collection were used:

• Focus group discussion

The method was used for data collection in different groups in the community. The groups include women, youth's community leaders and district extension officers who were involved in the project.

Group discussion Involved 10 to 12 people brought together in a single session of approximately an hour to generate ideas and suggest strategies. This method was helpful as project stakeholders were able to share their feelings and opinions about the project. It was also used to obtain in depth understanding of attitudes, impressions and insights (qualitative data) on variety of issues from the group.

The method was also friendly to group members as were able to discuss very open and give his or her opinion where possible. Focus group discussion also helped the participants to learn from each other the way the project implementation was done and participation of each parties.

Group discussion facilitator used five main guiding questions:

(a) Opening question: *Tell us your name and how long you have been participating in the program.*

(b) Introductory question: *what was it that you first learned about the program?*

(c) Transition Question: *Think back to when you first became involved with the program. What were your first impressions?*

(c) Key question: *In what way do you think your life will be different because of your participation on the program?*

(d) Ending question: *Is there anything we should have talked about, but didn't*

• Participant – observation

Some of the community members from different groups i.e. women, men, youths and leaders participated in observation of activities already done during the implementation of the project.

This method was suitable during the evaluation because community members were able to see physically what had already been completed and what was not.

It was also easy for the participants to recommend and give their opinion immediately on how to go about in order to complete intended project. During the observation, participants were taking notes on their note books on what was observed. Furthermore, the facilitator of the process (The author) used observation guideline that would simplify analysis of the observation. There were two major guiding questions that evaluation team required to consider.

The first question was; *what do you see is/are in place and what was there before the date of evaluation*. The second question was; *do you have any comments to what is observed*. The evaluation facilitator also use camera to take photos as evidence of what had been done during the project implementation. Sites visited by the observation team include, proposed distribution points sites, water sources currently on use. Another area visited is project equipment stores where all pipes and its accessories were kept.

• **Review of project records**

All documents related to project were reviewed. These included minutes of the meetings, communication between the organization and the donor and other stakeholders involved in the project implementation. The evaluation process was facilitated by the author. He used project diary to review some important information documented for the whole period of project implementation.

5.2.2 Data analysis and findings

Data analysis was done after review of all collected data basing on five key evaluation questions developed. Because evaluation process was participatory, sometimes analysis of the data was done during the discussion.

Summary findings are in monitoring and evaluation results in table 8. However, the following were the general answers based on the evaluation questions.

o *Did we do what we intend to?*

Respondents agreed that most of the activities planned to be done were done although some of them were not completed. They mentioned that construction of the scheme had not yet started although they were sure that it would start in April 2007 as pipes and other accessories had already been purchased and delivered to the site.

o **What did we learn about what was implemented and what was not implemented?**

During the discussions, respondents mentioned that, full participation of all stakeholders in development activities is the only way for success of community based development projects. This worked well from the initial stage of the project. However, it was revealed that neighbor communities were not involved in this project although they would be among the beneficiaries. This was not done well. Some members of the community and leaders who were involved in the evaluation process confirmed that there was participation and cooperation among the community members. Communities increased their trust to their leaders and thus they believed that they had enough capacity to manage the project and attain the expected results of increasing accessibility of water supply within the community in future. Full participation of the community members in the project was highly encouraged to proceed in order to achieve the intended objectives.

Communities were asked to participate more in cash and kind contributions, as it would be required. Attending meetings and transparency among the leaders during the project implementation made them different.

o **What do we plan to do with evaluation findings for continuous learning**

It was suggested that observations and comments given during the evaluation should be incorporated in the next phase of project implementation plan in order to have smooth implementation.

Moreover, the spirit which was shown by the communities especially in participation in both kind and cash contributions should be maintained for the success of the project.

5.2.3 Project Sustainability

The project established on the felt and priority need of the community. During the needs assessment, communities ranked water supply as priority. Therefore strategies were done to solve the problem and meet their desired condition of having sufficient water supply. This gives hopes that the project will stay longer as it is established from the real felt need of the communities of Tambukareli.

Communities through their leaders, have experience in management of the water source. Currently, the community owns one borehole, which is the only reliable water source for residents of Tambukareli community. There is a bank account and communities have to pay 10 shillings per 20 litres.

The money collected is sent into the bank account and later spent for operation and maintenance of the well. For the last 3 years, the well has successfully operated and it was repaired when major and minor breakdown occurred. Following this experience, there is hope that the same spirit will be applied in the new project because there is no big gap of ideas between the leaders and the users.

Capacity building has been imparted to the water user group leaders for the sake of increasing knowledge and skills. The aim is to enable leaders to manage and operate the project efficiently and effectively.

During the training various important areas were taught. These included simple bookkeeping and accounting, water tariffs, project rules, operation and maintenance, system operations legal and institutional framework.

The project has full support from the district authorities and other stakeholders including Plan International, a development organization. Geita District Council, in its next five years development plans, has put clearly aside the need to have access to water supply for its people in Geita town and its outskirt areas. In order to overcome the problem, the district has approached different donors including the central government for support. It has also encouraged the communities to have their own strategies for facing water shortage challenges in their areas. Thus, this project has been in line with district plans, and through the district water engineer it has the task of day-to-day supervision of the project.

CHAPTER SIX
CONCLUSION AND RECOMMENDATIONS

This chapter reviews the results of the work done. It reports objectives that were fully achieved and those not achieved. Factors or conditions that greatly affected the ability to complete the project and achieve all stated objectives are explained.

The chapter also provides experience gained during the project implementation. It recommends to others attempting similar project on strategies and best practices. Next steps for smooth implementation of this project are also explained here.

6.1   Results

The work done for one and a half years has been very successful to the extent that, it creates hope that in the next few months the goal and objectives of the communities will be achieved. It is successful work because the author has also managed to meet his objectives of facilitating the process at this point where the main project activities are expected to be completed in next the few months.

The Responsibility of the student was to offer technical assistance to Tambukareli water users group. He becomes a change agent at the disposal of the organization and concentrates in giving advice in various capacities, which was necessary and appropriate. During the needs assessment, communities identified the need to have water supply scheme, but how to secure funds to meet their objectives was the assignment given to the student as technical advisor.

It was later learnt that, in order to accomplish the community objective, a project proposal was needed in order to sell to interested donors or stakeholders. So the main service and objective of the student was to support the community in preparation of water supply project design, and conduct capacity building to the community leaders.

As stated in other chapters, preparation of the project proposal was successfully completed. Submitted and the donor released the funds for implementation. However, there are other associated works with this main service, which were also completed by the student. This include, organization registration process including preparation of organization constitution. All associated activities are ongoing in a very good progress. It is highly expected that the goals and objectives of the project will be met in the next few months after the project completion; successful completion of the project will mean that communities will have access to clean and safe water supply services closer to their user home. That means, it will contribute to the improvement of the health and living standards of the target communities.

It is also expected that the project will be operational with minimum supervision from the district officials. Water user group leaders, community members in general will have basic skills in management of water sources and water scheme management. Community management has become the framework for implementing water supply systems in rural areas in developing countries. It has yielded significant achievements, but it has not succeeded to supply water on a large scale and to secure long term sustainability of water supply systems. For that reason institutional support to community managed water systems is needed. That is the core and main objective of scaling up: indefinite sustainability (scaling up in time) and 100% coverage (scaling up in space) of community managed water supply systems.[16]

In summary, the capacity building objective is fully achieved. What remains is just follow up and support to leaders in order to meet their targeted plans. The second objective of improving accessibility to water supply has not yet been achieved.

[16] *IRC International Water and Sanitation Centre, Netherlands*

In order to meet it, construction work needs to start immediately as construction materials have already arrived. The delay of supplier to supply construction materials from Dar es Salaam has greatly affected timely completion of the project. This is unexpected occurrence noted during the implementation of the project.

6.2   Recommendations

- Based on the survey results, the following are the recommendations for next steps:
- In order to meet the intended objectives for improving accessibility of water supply in the study area, the option recommended by the communities (piped Water Supply scheme) is expensive. It needs a lot of money to complete.
- The communities themselves may not manage it in a short time. Communities should be facilitated to secure funds for the project from other interested partners including the government if we want to meet one of the national development targets of providing sufficient water supply closer to the user homes by the year 2025.
- Training program on water management and project management skills to the leaders of Tambukareli water users group is one of the important tools for project sustainability.
- The size and activities of the project need skilled and committed people who will operate and manage the whole process of project implementation. Therefore, leaders should make sure that all project personnel's are recruited, and trained to attain enough skills required.
- Community mobilization to participate in the process of project implementation should be strengthened through trainings, contributions in both kind and cash whenever necessary and regular collection of water tariffs,

in order to meet the intended objectives as well as sense of ownership and sustainability.

- Community participation and involvement of all stakeholders is of paramount importance in implementing this kind of project. People should own the process as well as the facility that will be built in order to achieve the expected objectives.

- Continuous capacity building to community in managing the facility and the organization will strengthen the sense of ownership and sustainability of the project. This is because, if communities are well trained in management of the water scheme, the water project can be operational without support from outside funding for daily service and maintenance.

- There is need to emphasize the multiple use of water of which communities can use for both domestic and economic activities. Communities should be encouraged to use water sources for any economic activities such as gardening and brick making for improving living standards of the families.

- As a water supply scheme is highly expensive, proper study/design and the use of the qualified water contractors and water consultants is recommended in construction of water supply scheme for the better result.

Bibliography

Arlene F (1985), How to conduct survey: *a step-by-step guide.* Sage Publications Inc. United Stated of America Blagbrouh, V (2003) How *Water Aid looked back;* Waterlines, Vol.22 No 1. pg. 35-38 London, UK

Carter, R.C et al (1993), Lessons *Learned from the UN Water Decade,* in Journal of intergraded water and environmental managers. London UK.

Carter R.C et al; (1999), Impact and sustainability of community water supply and sanitation programs in developing countries. *In journal of the chartered institution of Water and Environmental Management,* Vol.13 No 4 292-296. London UK

Carter R.C; (2004), Water and Poverty, will the dream of sustainable development be realized? *Second annual Cambridge lecture series in sustainable development. Centre for sustainable development department for engineering, Cambridge University, 10 March 2004 in developing countries*

Center for Development and Population Activities; (1994), Project Design for Program

Managers, Washington DC Council Health Management Team; (2004), Geita District Health Report 2004

71

Jennifer D. et al (2002), Taking Sustainable Rural Water Supply Service to scale. *A discussion paper. Washington USA DFID (2004), Departmental Report 2004': A department for International Development Policy Paper. UK, British Government DFID (2004), Water Action Plan: A department for International Development Policy Paper. UK, British Government.*

European Commission (1998). *Towards Sustainable Water resources management, a strategic Approach. Luxemburg, office for official Publications of the European communities.*

Jarman, J *et al* (1997), WAMMA: Empowerment in Practice. *London UK* Lockwood, H; (2002), Institutional support mechanism for community managed Rural Water Supply and Sanitation Systems in Latin America. Strategies *Report 6 for bureau for Latin America and the Caribbean.*

Ministry of Water & Water Aid (2002) Water and Sanitation in Tanzania - Report Most Clearinghouse, best Practices *(2002) Health through Sanitation and Water in Tanzania (HESAWA)*

Mathew, B (2005) Ensuring Sustained Beneficial outcomes for Water and Sanitation Programs in the Development world. IRC delft, the Netherlands.

Moriaty P. et al (2003) the productive use of domestic water supplies. How water supplies can play wider role in livelihood improvement and poverty reduction. *Conference proceedings: international symposium, water poverty and productive uses of water at the Household level 21-23 January2003, muldersdriff South Africa.*

Overseas Development Institute & Water and Sanitation Program Africa-Water Policy Briefing, No. 5 November 2004 Overseas Development Institute & Water Aid -Water policy Brief 3: July 2002.

Plan International Tanzania; (2003) CPME Baseline Report -Geita District, Country office, Dar es Salaam

Save the children; (1995) Save the children Development Manual No.5; Tool Kits. Save the children-London Schoutern, (2003). Community Water, Community Management, from system to service in Rural Areas. *London, IRC International Water and Sanitation Centre ITG*

Tanzania National Website (2005) The 21st Annual Water Experts Conference (AWEC) Report

Vice President's Office-Poverty Eradication *(2005)* National Strategy for Growth and Reduction of Poverty (NSGRP)

WHO/UNICEF, (2002), Global Water supply and Sanitation assessment 2000 report.

WSSCC (2000) Vision 21. A shared vision for hygiene, sanitation and Water supply and a framework for action. *Geneva, Switzerland, Water supply and Sanitation Collaborative Council*

WSSCC (2004), Resource pack on the water and sanitation millennium development goals. *Geneva, Switzerland, Water supply and Sanitation collaborative Council*

WSSCC, (2004) listening, to those working with communities in Africa, Asia and Latin America to achieve the UN goals for Water and Sanitation. *Geneva, Switzerland, Water supply and Sanitation collaborative Council.*

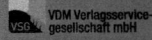

Printed by Books on Demand GmbH, Norderstedt / Germany